FLORA OF TROPICAL EAST AFRICA

MONTINIACEAE

B. Verdcourt*

Small trees or shrubs. Leaves alternate, opposite or subopposite, simple, entire, penninerved, usually petiolate, deciduous; stipules absent. Flowers dioecious in terminal or axillary inflorescences, the ♂ flowers in few-flowered panicles, the ♀ flowers solitary or rarely paired. Male flowers: calyx-tube cupular or flattened out, ± entire or shortly 3–5-lobed; petals yellow-green, 3–5, imbricate, slightly fleshy, deciduous, inserted below the margin of the fleshy disk; stamens yellowish, 3–5, similarly inserted and alternating with the petals; anther-thecae dehiscing by longitudinal slits; rudimentary ovary absent or minute. Female flowers: calyx-tube elongate, shortly produced beyond the ovary, scarcely or minutely 4–5-lobed at the apex; petals 4–5, imbricate, slightly fleshy, deciduous; disk fleshy, epigynous, 4–5-angled; staminodes 4–5, inserted below the margin of the disk and alternating with the petals; ovary inferior, (1–)2-locular; style short, thick, 2-fid or 2-lobed; stigmas 2, large; ovules (2–)3–9(–12) in each locule, axile, 1–2-seriate, erect or pendulous, anatropous. Fruit a loculicidally dehiscent capsule or indehiscent. Seeds ± globular or compressed and winged; endosperm copious or absent; embryo straight, with round compressed cotyledons.

A small family of two genera occurring in eastern tropical and southern Africa and Madagascar. Only *Grevea* occurs in the Flora area; the other genus, *Montinia*, containing the single species *M. caryophyllacea* Thunb., occurs in South Africa (Cape Province), South West Africa and just reaches Botswana. The affinities of the family are obscure; Mr. Milne-Redhead at first thought it best placed near the Oliniaceae but later considered it might, despite the inferior ovary, belong near to Celastraceae but this has been refuted by workers with detailed knowledge of the family. Hutchinson was certain that the reference to Escalloniaceae first made by Harvey in 1842 is correct but I am inclined to leave it near to Onagraceae and Oliniaceae where it at present resides in the Kew Herbarium. As has been pointed out by Milne-Redhead and is very evident from anatomical evidence** *Montinia* is not closely related to *Grevea* but these two genera are undoubtedly nearer to each other than to any other genus and I am quite agreed that they can be considered to form a small family on their own.

GREVEA

Baill. in Bull. Soc. Linn. Paris 1: 420, 477 (1884); Milne-Redh. in Hook., Ic. Pl. 36, sub t. 3541–4: 6 (1955); G.F.P. 2: 30 (1967)

Small trees or shrubs. Leaves opposite or subopposite. Young foliage buds in leaf-axils and at node-scars with multicellular hairs, appearing as a fringe to the scars. Male flowers in axillary inflorescences and ♀ flowers terminal, solitary; bracts very small or obsolete. Male flowers: calyx-tube cupular, subentire, exceeding the disk; petals 3; stamens (2–)3, with introrse anthers;

* This account is based to a large extent on the work carried out by Mr. E. Milne-Redhead and partly published in Hook., Ic. Pl. 36, t. 3541–4 (1955), where he established the new family accepted here.
** Dr. Metcalfe stated *in litt.* that he still found it very hard to believe that *Montinia* and *Grevea* belong to the same family, because the corresponding structure in the two genera is very different.

rudimentary ovary minute or ± absent. Female flowers: calyx-tube smooth and striate or tuberculate, with (3–)4 inconspicuous incurved lobes; petals 4(?–5); staminodes 4; style thick, 2-lobed at the apex; each lobe covered entirely by the stigmatic surface on its inner face so that the surface is continuous, but only on the margins and outer areas of the outer faces (the stigmas thus appearing to be 2), ± ovoid with a part of each style-lobe left in the middle of the outer faces; each lobe may again be bilobulate or (*fide* Capuron) the style may be bifid, each branch bearing a bilobulate stigma or even the style divided to the base and the stigmatic zones extending towards the base on the inner side of the branches. Ovary completely 2-locular or sometimes 1-locular, the 2 parietal placentas meeting at the centre but not completely joining. Ovules 2–9(–12 *fide* Capuron) on each placenta, pendulous, 1–2-seriate. Fruit indehiscent, with coriaceous to crustaceous pericarp, crowned by the calyx-limb, the placentas expanding to form 1–2 masses (according to whether the ovary is 1–2-locular), but eventually the mass of seeds embedded in the fleshy endocarp forms a single mass in a unilocular fruit. Seeds ovoid or globose, 2–9(?–12) to each placenta; testa consisting of a single layer of polygonal cells, membranous, obscurely finely reticular; endosperm abundant, horny.

A genus of 2 species occurring in western and north-western Madagascar and eastern tropical Africa, one with 2 subspecies, the other with 3.

Fruits and developing ovaries almost smooth, with
    no more than rows of low nodules*   .   .   1. *G. madagascariensis*
Fruit and developing ovaries tuberculate to
    distinctly echinate   .   .   .   .   .   2. *G. eggelingii*

1. **G. madagascariensis** *Baill.* in Bull. Soc. Linn. Paris 1: 420, 477 (1884); Erdtman, Pollen Morph.: 400 (1952); Milne-Redh. in Hook., Ic. Pl. 36, sub t. 3541–4: 7 (1955); Capuron in Adansonia, sér. 2, 9: 512 (1969). Type: Madagascar, Morondava, *Grevé* 249 (P, holo., K, photo.! & iso.!)

Shrub to medium-sized tree usually 2–3 m. tall; older stems very pale brownish or greyish white, minutely longitudinally sulcate, densely covered with lenticels, the lower nodes bearing prominent leaf-scars; young stems green with white lenticels; young buds with short obscure to quite distinct woolly hairs. Leaves opposite; blades elliptic to ovate-elliptic or almost round, 4–19(–25) cm. long, 1·6–8·5(–20) cm. wide, narrowly acuminate or rarely rounded at the apex, cuneate to rounded or in large leaves rarely even distinctly cordate at the base, glabrous; lateral nerves 4–5 pairs; petiole 0·7–4 cm. long. Male inflorescences cymose, 6–12-flowered; peduncle 1·2–3·8 cm. long; pedicels 3–9 mm. long. Male flowers: calyx 1–2 mm. long; petals oblong to oblong-spathulate, 3–4·2 mm. long, 2–3 mm. wide, rounded, slightly narrowed to the base, often slightly erose; stamens with filaments 1·2–3 mm. long, anthers 2–2·2 mm. long; rudimentary ovary obsolete or 0·5 mm. long. Female flowers: calyx-tube 1·2 cm. long, slightly expanded below, cylindrical above, either smooth or with obscure nodules in longitudinal stripes; pedicel-like stipe 3–7(–9) mm. long; lobes broadly triangular, 0·5–1 mm. long; petals oblong-spathulate, 4·8 mm. long, 2·2 mm. wide, rounded at the apex, slightly narrowed to the base, reflexed, slightly erose; staminodes with filament part 2 mm. long and triangular anther part 1·2 mm. long; style 4–5 mm. long; stigma 1·5–1·8 mm. long. Fruit flask-shaped, 3·2–4 cm. long, 1–1·2 cm. wide, either brownish, obscurely ribbed and faintly rugulose or greyish and with distinct nodulated ribs, crowned with the

* Since the male of *G. madagascariensis* subsp. *keniensis* is unknown, no distinction for males can be given, but geography alone will sort out the three taxa involved.

persistent style. Seeds reddish brown, 4–16(–18), or *fide* Capuron even –24, ellipsoid-subglobose, 5·5 mm. long, 4·5 mm. thick, slightly compressed, finely shallowly reticulate.

subsp. **keniensis** *Verdc.* in K.B. 28: 147 (1973). Type: Kenya, Diani, *Napier* 3348 (EA, holo.!, K, iso.*)

Buds in axils of leaves and on old nodal scars distinctly hairy. Leaf-blades ovate-elliptic, 7–19 cm. long, 3–8 cm. wide, prominently narrowly acuminate at the apex. cuneate at the base. Immature fruit 2·6 cm. long, 1·1 cm. wide, the beak short, 3 mm, long, the stipe 3 mm. long.

KENYA. Kwale District: Diani, 1 June 1934, *Napier* 3348!
DISTR. **K7**; not known elsewhere
HAB. Evergreen forest; near sea-level

SYN. *G. sp. A* sensu Milne-Redh. in Hook., Ic. Pl. 36, sub t. 3541–4: 9 (1955)

NOTE. Despite intensive searches no further material of this has been discovered but it is unlikely to be extinct.

DISTR. (of species as a whole). Subspecies *madagascariensis* and *sublevis* Verdc. occur in Madagascar.

2. **G. eggelingii** *Milne-Redh.* in Hook., Ic. Pl. 36, sub t. 3541–4: 7 & t. 3543–4 (1955); Mendes in Bol. Soc. Brot., sér. 2, 44: 299, fig. 1/C (1970). Type: Tanganyika, Morogoro R. below Bahati, *Eggeling* 6438 (K, holo.!, B, BR, EA, G, P, TFD, iso.!)

Shrub or small tree (1–)3·5–7·5 m. tall, often thicket-forming by means of sucker shoots; older stems pale grey-brown or greyish white, brittle, minutely longitudinally sulcate, densely covered with small lenticels, the lower nodes bearing obcordate leaf-scars; young stems green with white lenticels; young buds in leaf-axils and at leafless nodes woolly. Leaves opposite or sub-opposite; blades elliptic or ovate, (1·3–)3·5–23·5 cm. long, (0·6–)1·4–14 cm. wide, shortly acutely acuminate at the apex, narrowly to broadly cuneate at the base, the margin entire but sometimes slightly undulate, glabrous; lateral nerves 4–5 pairs, arcuate; petiole 0·4–3(–5) cm. long. Male inflorescences ± umbel-like cymes, up to 8-flowered, in the axils of juvenile leaves; peduncle 0·4–5 cm. long; pedicels 0·2–1·8 cm. long. Male flowers: calyx ± 1–2 mm. tall; petals white or yellow-green, oblong or oblong-subspathulate, 4–5 mm. long, 2·5–3 mm. wide, rounded at the apex, slightly narrowed to the base, strongly reflexed, very slightly erose; stamens with filaments 1·2–1·5(–3) mm. long; anthers (1·6–)2·5–3 mm. long; rudimentary ovary 0·9–1·5 mm. long. Female flowers: calyx-tube 8 mm. long, slightly expanded below, cylindrical above, irregularly tuberculate, pedicel-like stipe 1–4(–8) mm. long; lobes triangular, ± 0·5 mm. long; petals creamy, oblong, 3·5–4 mm. long, 2·5 mm. wide, rounded at the apex, slightly narrowed at the base, strongly reflexed, very slightly erose; staminodes ± 1 mm. long; style white, thick, 3–4(–6) mm. long, shortly 2(–3)-fid at the apex, crowned with the ovately cap-shaped stigmas 1·8–2·5 mm. long. Fruit flask-shaped, 1·5–3(–3·5) cm. long, 1–1·6 cm. wide, irregularly densely spinous-tuberculate or echinate and with smaller rounded tubercles on the widened part, the neck ± smooth, crowned with the persistent style. Seeds reddish or yellowish brown, (4–)5–11(–12), ± round, compressed, ± 5–6 mm. long, 5 mm. wide and 2·5–3·5 mm. thick.

* The Kew sheet of this gathering has been mislaid since long before Milne-Redhead's paper was written. It had been variously placed in Rubiaceae and Oliniaceae and removed to another family, but the botanist concerned has forgotten where he put it!

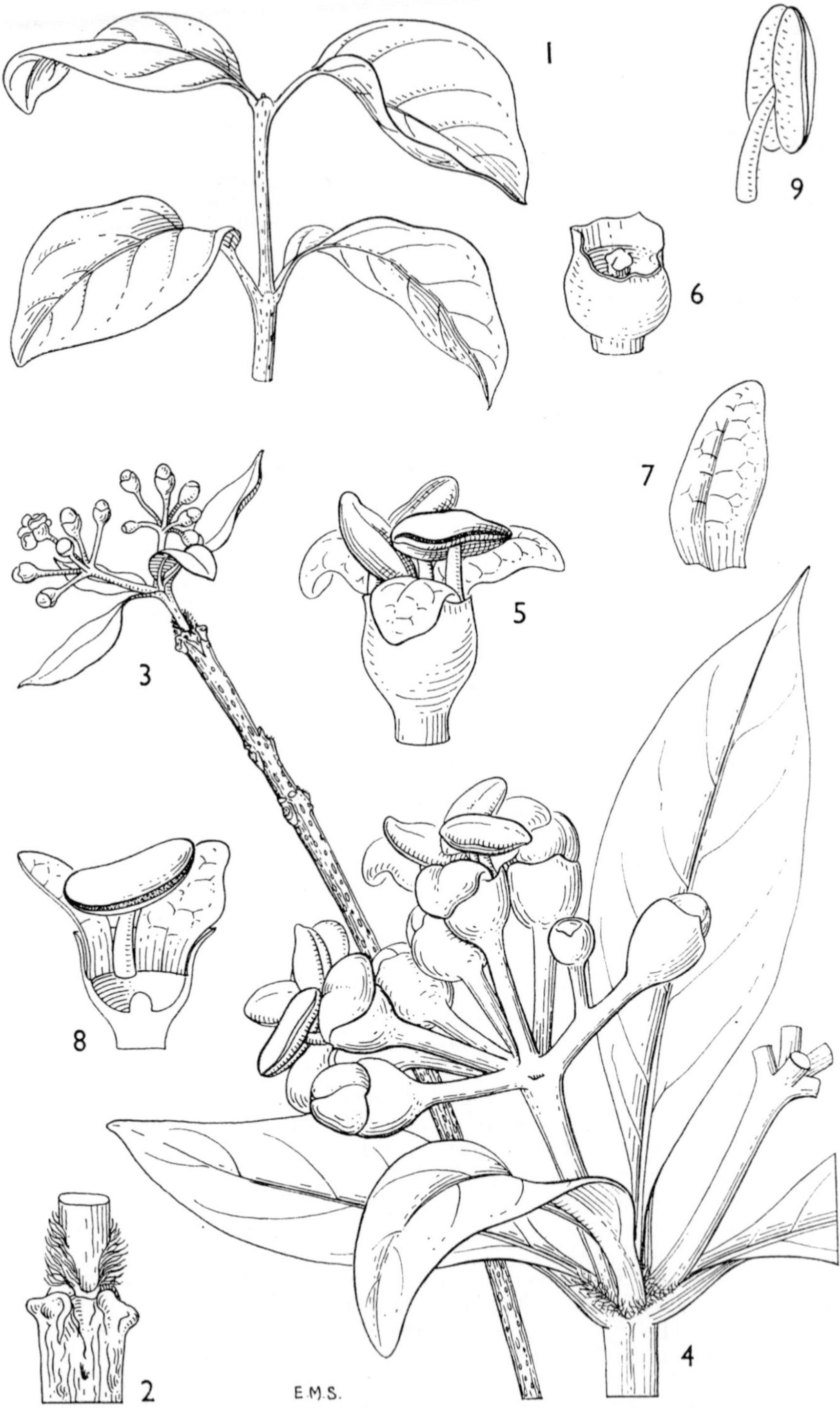

E.M.S.

FIG. 1. *GREVEA EGGELINGII* subsp. *EGGELINGII*, male—**1**, sterile branchlet, × ⅔; **2**, node of stem, showing pubescence covering the buds, × 4; **3**, flowering branch, × 1½; **4**, leaves and inflorescence, × 4; **5**, flower, × 6; **6**, calyx, partly removed to show pistillode, × 6; **7**, petal, × 6; **8**, longitudinal section of flower, × 6; **9**, stamen, × 6. Mostly from *Eggeling* 6442. Drawn by Miss E. M. Stones. Reproduced by permission of the Bentham-Moxon Trustees.

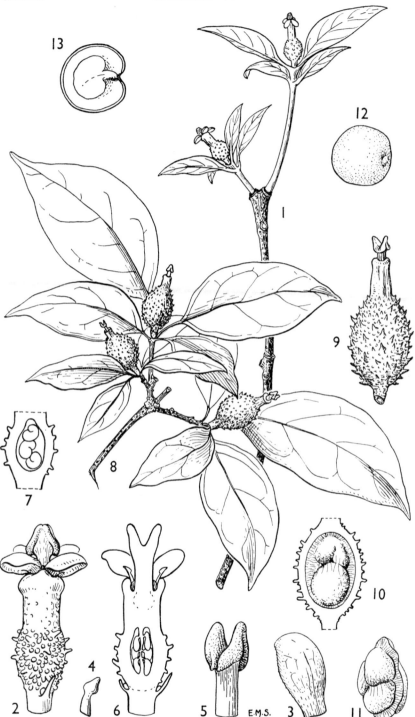

FIG. 2. *GREVEA EGGELINGII* subsp. *EGGELINGII*, female—**1,** flowering branchlet, × 1; **2,** flower, × 4; **3,** petal, × 6; **4,** staminode, × 8; **5,** style and stigma, × 6; **6,** longitudinal section of flower, × 4; **7,** ovary, with wall removed to show ovules (diagrammatic), × 4; **8,** fruiting branchlet, × ⅔; **9,** fruit, × 1½; **10,** fruit cut open to show septum and seeds, × 1½; **11,** seeds, covered with endocarp, adhering to septum, × 1½; **12,** seed, × 3; **13,** seed, split open to show embryo, × 3. 1–7, from *Eggeling* 6441; 8–13, from *Eggeling* 6439. Drawn by Miss E. M. Stones. Reproduced by permission of the Bentham-Moxon Trustees.

subsp. **eggelingii**

Male inflorescences with peduncles 1·5–2 cm. long; pedicels 2–7(–8) mm. long; filaments mostly 1·2–1·5 mm. long. Fruits with lower part densely tuberculate, the tubercles slender and sharply pointed, 1–2(–3) mm. long; ovules and seeds 2–3(–4) per locule. Figs. 1 & 2.

TANGANYIKA. Handeni District: Tamota, July 1950, *Semsei* 613 !; Morogoro District: Mtibwa Forest Reserve, Dec. 1953, *Semsei* 1514 ! & Magandu Public Land, June 1953, *Semsei* 1274 ! & Morogoro R., below Bahati, Dec. 1952, *Eggeling* 6439 !
DISTR. T3, 6; not known elsewhere
HAB. Dry riverine thicket, *Brachystegia* woodland, riverine forest, often on steep well-drained slopes; 420–600 m.

SYN. *G. sp.* sensu Erdtman in Webbia 11: 408 (1955)

NOTE. *Renvoize & Abdallah* 1832 (Tanganyika, Kilosa District, Mikumi National Park, 30 Apr. 1968) has the fruit-spines longer than in the type thus approaching the next subspecies. In the one fruit investigated one placenta bore 3 seeds and the other 4 thus confirming this intermediate position.

subsp. **echinocarpa** (*Mendes*) *Verdc.* in K.B. 28: 149 (1973). Type: Mozambique, Cabo Delgado, between Montepuez and Nantulo, *Torre & Paiva* 11708 (LISC, holo. !, COI, K, P, iso. !)

Male inflorescences with peduncles (2–)3–5 cm. long; pedicels 0·7–1·8 cm. long; filaments mostly 2·5–3 mm. long. Fruits with lower part densely echinate, the spines stouter, 4–6(–8) mm. long; ovules and seeds 5–6 per locule.

TANGANYIKA. Mikindani District: Mahurunga, 5 Mar. 1964, *Mason* 1 ! (♂) & same area, ? Mar. 1965, *Mason* in Kew H. 1255/65 ! (♂ & ♀) & same area, ? Jan. 1967, *Mason* in Kew H. 739/67 ! (♂)
DISTR. T8; Mozambique
HAB. Forest and forest edges in damp sandy bottoms of valleys feeding the Rovuma R.; ? under 60–450 m.

SYN. *G. eggelingii* Milne-Redh. var. *echinocarpa* Mendes in Bol. Soc. Brot., sér. 2, 44: 299, fig. 1/A, B (1970)

NOTE. One collection of Mason's (*Kew H.* 151/61 from the same general area*) has the peduncles of the male inflorescences 1–1·5 cm. long and the pedicels 3 mm. long but they are not mature. Nevertheless it seems unlikely that they will ever reach the dimensions given above. Information on the development of the inflorescences is required. This collection together with that of Renvoize and Abdallah previously mentioned have influenced me in giving the plant subspecific rather than specific rank.

---

* I have since ascertained from Mr. P. Mason that one lot of material was collected on the Rondo Plateau as far as he can remember which at an altitude of some 600 m. has a different vegetation from the coast. It seems likely that this one aberrant lot came from there and might come closer to the typical subspecies. Only further well-localized collections will solve the problem.

INDEX TO MONTINIACEAE